DIFFERENT TYPES OF DINOSAURS YOU NEED TO KNOW

PHYSICAL FEATURES, DIETS AND IMPORTANT FACTS ABOUT THEM

DINOSAUR'S WORLD

Copyrights© Dinosaur's World
All Right Reserved

TABLE OF CONTENTS

Introduction

Chapter One

Tyrannosaurus Rex

    Physical Features and Characteristics

    Diet

    Some Important Facts about the T. Rex

Chapter Two

Triceratops

    Physical Features and Characteristics

    Diets

    Some Important Facts about Triceratops

Chapter Three

Velociraptor

    Physical Features and Characteristics

    Diet

Chapter Four

Stegosaurus

    Physical Features and Characteristics

    Diet

    Other Interesting Facts about Stegosaurus

Chapter Five

Brachiosaurus

   Physical Features and Characteristics

   Diet

   Some Interesting Facts about Brachiosaurus

Chapter Six

Apatosaurus

   Unique Features

   Diet

   Other Facts about Apatosaurus

Chapter Seven

Pteranodon

   Physical Features

   Diet

   Interesting facts about Pteranodons

Chapter Eight

Spinosaurus

   Physical Features

   Diet

   Interesting Facts about Spinosaurus

Chapter Nine

Allosaurus
- Physical Features
- Diet
- Interesting Facts about Allosaurus

Chapter Ten

Diplodocus
- Physical Features
- Diet
- Interesting Facts about Diplodocus

Chapter Eleven

Parasaurolophus
- Physical Features
- Diet
- Facts about Parasaurolophus

Chapter Twelve

Ankylosaurus
- Physical Features
- Diet
- Other Facts about Ankylosaurus

Chapter Thirteen

Iguanodon

    Physical Features

    Diet

    Further Facts about Iguanodon

Chapter Fourteen

Megalosaurus

    Physical Features

    Diet

    Further Facts about Megalosaurus

Chapter Fifteen

Camarasaurus

    Physical Features

    Diet

    Other Facts about Camarasaurus

Chapter Sixteen

Corythosaurus

    Features

    Diet

    Other Interesting Facts about Corythosaurus

Chapter Seventeen

Carnotaurus

    Physical Features

Diet

Other Interesting Facts about Carnotaurus

Chapter Eighteen

Majungasaurus

   Physical Features

   Diet

   Other Interesting Facts about Majungasaurus

Chapter Nineteen

Baryonyx

   Other Physical Features

   Major Diet

   Other Interesting Facts about Baryonyx

Chapter Twenty

Giganotosaurus

   Physical Features

   Diet

   Other Intriguing Aspects about Giganotosaurus

   Giganotosaurus and T. Rex Comparison

Conclusion

Bibliography and Further Reading

Introduction

Ever wondered what the giants that roamed the Earth millions of years ago looked like? Buckle up, junior paleontologists! This book will introduce you to some of the most fascinating dinosaurs that ever lived during the Mesozoic Era.

Note that the Mesozoic Era, also known as the Age of Dinosaurs, spanned a vast period of time from roughly 252 million to 66 million years ago. It marked the second era in the Earth's three-part geologic timescale, following the Paleozoic Era and preceding the Cenozoic Era. The word "Mesozoic" comes from the Greek words mesos (meaning middle) and zoe (meaning life), signifying its position between the Paleozoic ("ancient life") and Cenozoic ("recent life") eras.

The Mesozoic Era was a time of significant change for Earth. The supercontinent Pangaea gradually broke apart into the continents we knew today and the climate fluctuated between hot and humid to warm and dry. However, the era is most famous for the rise and reign of dinosaurs, a diverse group of reptiles that dominated the planet for millions of years.

The Mesozoic Era is divided into three distinct periods: Triassic, Jurassic, and Cretaceous. Each period witnessed unique developments in climate, geography, and the evolution of life. Here's a brief overview of each:

- **Triassic Period (252-201 million years ago):** The earliest Mesozoic period, characterized by the recovery of life following the Permian-Triassic extinction event. The climate was hot and dry, with vast desert landscapes. Dinosaurs emerged during this time, though they were small and relatively insignificant compared to their later kin.

- **Jurassic Period (201-145 million years ago):** Often referred to as the "Golden Age of Dinosaurs." The climate was warm and humid, with lush forests and abundant vegetation. Dinosaurs diversified tremendously during this period, giving rise to the iconic giants like Stegosaurus and Brachiosaurus.
- **Cretaceous Period (145-66 million years ago):** The final period of the Mesozoic Era, marked by a continuation of the warm climate and the reign of advanced dinosaurs like Tyrannosaurus Rex and Triceratops. Flowering plants emerged during this period, and the continents continued to drift further apart. The era ended with the Cretaceous-Paleogene extinction event, which wiped out non-avian dinosaurs and many other marine and terrestrial species.

The Dinosaurs were the dominant terrestrial vertebrates of the Mesozoic Era, ruling the land for over 180 million years. They were a diverse group of reptiles, ranging in size from the tiny Compsognathus, which weighed about as much as a chicken, to the gigantic Argentinosaurus, which could have weighed as much as 100 tons. Dinosaurs were bipedal (walked on two legs) or quadrupedal (walked on four legs), and many sported feathers or feather-like structures. They were classified into two main groups: saurischia (lizard-hipped) and ornithischia (bird-hipped). Saurischians included the giant sauropods, the fearsome theropods (including Tyrannosaurus Rex), and the prosauropods. Ornithischians included the armored thyreophores (like Stegosaurus), the ceratopsians (like Triceratops), and the ornithopods (like Iguanodon).

Note that paleontologists believe that birds actually evolved from theropod dinosaurs during the

Jurassic period. This connection is evident in many dinosaur fossils that show features like feathers and hollow bones, both characteristic of birds. The extinction of non-avian dinosaurs at the end of the Cretaceous period remains a subject of debate, but the most widely accepted theory involves a massive asteroid impact that caused a global climate catastrophe.

Paleontology is the scientific study of ancient life, from the earliest single-celled organisms to the dinosaurs that roamed the Earth millions of years ago. Paleontologists use fossils—the preserved remains of plants, animals, and other organisms—to piece together the history of life on Earth. Fossils can be bones, teeth, shells, footprints, or even traces of DNA. By studying fossils, paleontologists can learn about the evolution of life, the environments that existed in the past, and the mass extinction events that have shaped our planet.

Paleontology is a field that draws on many other scientific disciplines, including biology, chemistry, geology, and physics. Paleontologists use a variety of techniques to study fossils, including excavation, preparation, and analysis. Excavation involves carefully removing fossils from the rock or sediment in which they are buried. Preparation involves cleaning and repairing fossils so that they can be studied in more detail. Analysis involves using a variety of techniques to learn about the anatomy, physiology, and ecology of the organisms that the fossils represent.

The study of paleontology has led to many important discoveries about the history of life on Earth. For example, paleontologists have shown that dinosaurs were not cold-blooded reptiles, but rather warm-blooded animals that were closely related to birds. They have also shown that the Earth has undergone several periods of mass extinction, in which a large percentage of species have died out in a relatively short period of time.

In the following chapters, we will explore Dinosaurs incredible physical features, from razor-sharp claws and giant horns to long necks and spiky plates. We will also discover what they munched on, whether they were plant-eaters or meat-eaters. And along the way, we'll learn some cool facts about these prehistoric superstars!

So, are you ready to travel back in time and meet the dinosaurs? Let's get started!

Chapter One

Tyrannosaurus Rex

Imagine a giant lizard the size of a bus, with teeth as sharp as knives and a roar that could shake the ground. That is Tyrannosaurus Rex, also known as T. Rex for short! It lived millions of years ago and was one of the biggest meat-eating dinosaurs ever to walk the Earth.

Tyrannosaurus Rex, also known as the "king of the tyrant lizards," has captivated imaginations for over a century. But its story began much earlier, in the dusty plains of South Dakota. In 1900, paleontologist Barnum Brown unearthed partial vertebrae and a massive hip bone. These belonged to a creature unlike any seen before, hinting at a gigantic predator.

Brown, funded by the American Museum of Natural History, embarked on a relentless quest for more.

Over the next few years, meticulous excavations yielded a treasure trove of bones – a testament to T. Rex's fearsome presence. The scientific community was astounded. The discovery of Tyrannosaurus Rex marked a turning point in our understanding of prehistoric giants and continues to fuel our fascination with the lost world of dinosaurs.

Tyrannosaurus Rex

Physical Features and Characteristics

T. Rex was a fearsome predator, but what exactly made it so terrifying?

Let's dissect its incredible features:

i. **Size:** T. Rex could grow up to 40 feet long and weigh up to 18 tons – that is heavier than several elephants combined!

ii. **Head:** Its skull was massive and powerful, filled with huge, sharp teeth that were nearly a foot long. These teeth were perfect for puncturing and ripping flesh.

iii. **Jaws:** T. Rex's jaws were incredibly strong, with a bite force that could crush even the toughest bones.

iv. **Arms:** Although they were short compared to its body, T. Rex's arms were surprisingly strong and muscular. They had two clawed fingers on each hand, which may have been

v. **Legs:** Its powerful legs helped T. Rex run surprisingly fast for its size – estimates suggest speeds of up to 20 miles per hour in short bursts.

Diet

Now, let us move on to what this mighty hunter liked to eat. T. Rex was a carnivore, which means it ate meat. Some of T. Rex's favorite meals includes:

a. **Hadrosaurs:** These duck-billed dinosaurs were plentiful herbivores and perfect targets for T. Rex with their thick, fleshy bodies.

b. **Ceratopsians:** Triceratops, with their bony frills and horns, might seem tough, but T. Rex could have attacked them in groups or targeted the young or weak.

c. **Ankylosaurs:** These armored dinosaurs were a challenge, but T. Rex's powerful bite is strong enough to pierce their armor.

Some Important Facts about the T. Rex

Some other cool facts about T. Rex Include:
- T. Rex had eyes positioned forward on its head, giving it good depth perception to track prey.
- Although not the smartest dinosaur, T. Rex had a larger brain than most meat-eaters of its time.
- Scientists believe T. Rex may have been aggressive and possibly even scavenged for food sometimes.

Therefore, T. Rex was a truly magnificent creature, and even though it lived millions of years ago, it continues to capture our imagination.

Chapter Two

Triceratops

Next up in our prehistoric parade is the iconic Triceratops! This herbivore, easily recognized by its three horns and bony frill, was a fascinating creature that roamed the Earth alongside T. Rex.

The iconic Triceratops wasn't always recognized for the three-horned wonder it is today. The first clue emerged in 1887, unearthed by a man named George Lyman Cannon near Denver, Colorado. What he found was a partial skull unlike anything documented – a pair of brow horns fused to a bony frill.

Mistaken for a giant bison at first, it was paleontologist Othniel Charles Marsh who recognized the true nature of the discovery. By 1889, more Triceratops remains came to light, revealing the distinctive three horns and massive bony frill. This landmark discovery unveiled a bizarre and magnificent herbivore that roamed the landscapes of Laramidia, the western continent of North America, during the Late Cretaceous period.

Triceratops

Physical Features and Characteristics

Its unique features include:

i. **Three Horns:** The most striking feature of Triceratops is its three horns. The two larger horns grew above the eyes and could be up to three feet long! The smaller horn sported a bump near the snout. Scientists believe these horns were primarily for defense against predators.

ii. **Frill:** The bony frill extending from the Triceratops' neck served multiple purposes. It protected the neck, provided attachment points for powerful neck muscles, and might have even helped regulate body temperature.

iii. **Beak:** Triceratops had a strong beak perfect for ripping and tearing through tough plants.

iv. **Size:** While not as massive as T. Rex, Triceratops was still a formidable sight, reaching up to 30 feet long and weighing up to 12 tons.

v. **Four legs:** Triceratops walked on all fours, with powerful hind legs for support and strong front legs for maneuvering and browsing.

Diets

Unlike T. Rex, Triceratops was a plant-eater, munching on a variety of vegetation such as:

a. **Ferns:** These formed a large part of the Triceratops' diet, providing essential nutrients.

b. **Cycads:** These cone-bearing plants were another staple food source.

c. **Angiosperms:** Early flowering plants have also been on the Triceratops' menu.

Note that the Triceratops lived in herds for protection and easier access to food.

Some Important Facts about Triceratops

Some additional interesting facts about these three-horned giants include:

- Triceratops had a keen sense of smell to locate food sources and detect predators
- Similar to T. Rex, Triceratops' eyes were positioned forward, giving them depth of perception and ability to better navigate their surroundings.
- Estimates suggest Triceratops could live up to 60 years!

Thus, Triceratops used to be magnificent herbivores that played an important role in their ancient ecosystems and their unique features continue to spark our curiosity about the prehistoric world.

Chapter Three

Velociraptor

Now we meet the Velociraptor, a dinosaur that might be smaller than depicted in movies, but just as amazing! This raptor was a cunning predator with a whole arsenal of adaptations for hunting.

The velociraptor's path to fame began not with a fearsome skeleton, but with a single claw. In 1922, amidst the Gobi Desert of Mongolia, paleontologist Henry Nicholas discovered a curved, razor-sharp claw unlike any known animal. It belonged to a creature he named Velociraptor, meaning "swift seizer."

For decades, the true picture of the velociraptor remained elusive, with only fragmentary remains scattered across the Gobi. Then, in 1971, came a groundbreaking discovery. A joint Mongolian-Polish expedition unearthed a stunning fossil: a velociraptor locked in an epic final struggle with a Protoceratops. This exquisitely preserved specimen provided a wealth of information about velociraptor anatomy and behavior, solidifying its place as one of the most fascinating dinosaurs.

Velociraptor

Physical Features and Characteristics

Let's explore some features that made them so successful:

i. **Size:** Velociraptors were relatively small Theropods, measuring around 4-6 feet tall and weighing up to 40 kilograms. Their size, however, made them agile and stealthy hunters.

ii. **Sharp claws:** Velociraptor's most iconic feature is its enlarged sickle claw on each hind foot. This razor-sharp claw could slash prey and deliver a powerful killing blow.

iii. **Jaws and teeth:** Their jaws were lined with sharp, serrated teeth perfect for tearing flesh.

iv. **Intelligence:** Velociraptors were likely one of the most intelligent dinosaurs. Their brains, relative to their size, were quite large, suggesting they were cunning predators.

v. **Binocular vision:** Like T. Rex and Triceratops, Velociraptors had eyes positioned forward on their heads, giving them depth and balanced perception to track and target prey.

vi. **Pack hunting:** While some evidence suggests they might have hunted alone, Velociraptors may have also coordinated attacks in packs, making them even more formidable hunters.

Diet

What did these clever hunters eat? Velociraptors were carnivores and their diet likely consisted of:

a. **Smaller Animals:** Such as, Lizards, small ornithischians, and even baby dinosaurs could have been on the menu.

b. **Mammals:** Early mammals that shared their habitat likely served as prey.

c. **_Scavenging:_** Velociraptors may have also scavenged on carcasses of larger animals.

Some other interesting facts about Velociraptors include:

- Recent fossil discoveries suggest Velociraptors, and many other raptors, may have been covered in feathers!
- While not the fastest dinosaurs, Velociraptors were likely agile and swift runners.
- The large brain size of Velociraptors hints at their intelligence and possibly even some level of curiosity about their surroundings.

Thus Velociraptors were fascinating predators that played a crucial role in their prehistoric ecosystem. Their intelligence, combined with their physical adaptations, made them successful hunters for millions of years.

Chapter Four

Stegosaurus

Up next is the Stegosaurus which is an herbivore famous for its spiky plates and bumpy armor. This unique dinosaur was a gentle giant that stood out from the crowd during the Jurassic period.

The stegosaurus, with its distinctive bony plates and spiked tail, has been a source of fascination since its discovery in the late 19th century. The first clue emerged in 1877, when Othniel Charles Marsh, a paleontologist locked in a fierce rivalry with Edward Drinker Cope (the "Bone Wars"), unearthed some unusual vertebrae in Wyoming. These spiky bones hinted at a creature unlike anything previously encountered.

Marsh's relentless pursuit for more fossils paid off. In 1879, he unearthed a partial skeleton near Como Bluff, Wyoming, which he named Stegosaurus armatus, meaning "roofed lizard with armor." This discovery marked a significant step in understanding these bizarre herbivores that roamed North America during the Late Jurassic period.

Stegosaurus

Physical Features and Characteristics

Let's explore what made them special:

i. **Plates:** The most distinctive feature of Stegosaurus is its row of diamond-shaped plates running down its back. These plates were made of bone and may have served multiple purposes, including defense, temperature regulation, and display.

ii. **Spikes:** Along its tail, Stegosaurus sported sharp spikes that could be used to ward off predators. These spikes were particularly dangerous as Stegosaurus could swing its tail with great force.

iii. **Small head:** Stegosaurus had a relatively small head for its size, with a beak-like mouth perfect for munching on plants.

iv. **Bumpy body:** Along with the plates, Stegosaurus also had bony bumps running along its back and sides. These bumps

v. **Size:** Stegosaurus was a large dinosaur, reaching up to 30 feet long and weighing around 4 tons. Their large size, combined with their armor, helped them deter predators.

Diet

Now, let's see what this armored tank liked to eat:

a. **Low-lying plants:** Stegosaurus likely spent most of their day grazing on ferns, cycads, and other low-growing plants.

b. **Conifers:** These cone-bearing plants might have also been part of their diet. Their strong, small, flat grinding teeth means that Stegosaurus were well-suited for grinding up tough plant material.

Other Interesting Facts about Stegosaurus

Some other interesting facts about Stegosaurus include:

- Some theories suggest that the plates on Stegosaurus' back helped regulate their body temperature by absorbing sunlight.
- Stegosaurus' brain was relatively small compared to its body size, suggesting they weren't the most intelligent dinosaurs.
- Some scientists believe Stegosaurus may have lived in herds for protection and easier access to food sources.

Thus Stegosaurus were unique herbivores that played an important role in their ancient ecosystems. Their spiky armor and interesting physical features continue to capture our imagination today.

Chapter Five

Brachiosaurus

Imagine a giraffe on steroids; that is the Brachiosaurus!

This gentle giant was the tallest land animal ever to walk the Earth. Its long neck allowed it to reach leaves most other dinosaurs couldn't.

The story of the Brachiosaurus begins in the canyons of western Colorado, where a massive bone hinted at a giant unlike any other. In 1900, paleontologist Elmer Riggs unearthed a colossal shoulder blade while exploring the Morrison Formation. This single bone, much larger than anything from known dinosaurs, sparked a scientific hunt.

Riggs, funded by the American Museum of Natural History in New York, meticulously combed the area. Over the next few years, his team unearthed more bones, revealing an astonishing creature with a giraffe-like neck and incredibly long forelimbs. By 1903, enough evidence was gathered to introduce the world to the Brachiosaurus, the "arm lizard," a name reflecting its most prominent feature. This discovery not only unveiled a new dinosaur but also challenged existing ideas about size and diversity in the prehistoric world.

Brachiosaurus

Physical Features and Characteristics

Let us delve into its incredible features:

i. **Immense size:** Brachiosaurus was a behemoth, reaching up to 85 feet tall – that's taller than a six-story building! They could also grow up to 72 feet long and weigh an astounding 80 tons, heavier than several elephants combined.

ii. **Extra-long neck:** The most striking feature of Brachiosaurus is its incredibly long neck, which could make up almost half its entire body length! This adaptation allowed them to browse leaves from the tops of tall trees, where most other dinosaurs couldn't reach.

iii. **Small head:** Despite their massive size, Brachiosaurus had a relatively small head with a narrow snout and leaf-shaped teeth.

iv. **Powerful legs:** Even though their front legs were longer than their back legs, Brachiosaurus had strong legs for supporting its enormous body weight.

v. **Large nostrils:** Located on top of the head, the wide nostrils of Brachiosaurus may have helped them improve their sense of smell and possibly regulate body temperature.

Diet

Brachiosaurus was an herbivore, and its long neck was perfectly suited for its plant-based diet such as:

- **High leaves:** Their long necks allowed them to feast on leaves from the highest branches of trees, including conifers, cycads, and early flowering plants.

- ***Massive appetite:*** To sustain their giant bodies, Brachiosaurus needed to eat a tremendous amount of plant material every day.

Some Interesting Facts about Brachiosaurus

Here are some other interesting facts about Brachiosaurus:
- Due to their plant-eating diet and lack of sharp claws or teeth, Brachiosaurus were widely believed to be gentle giants.
- To circulate blood all the way up their long necks, Brachiosaurus happened to have a powerful heart and complex circulatory system.
- Some scientists believe Brachiosaurus may have lived in herds, possibly for protection and easier access to food sources.

In a nutshell, Brachiosaurus were truly magnificent creatures, and their reign as the tallest land animal is a testament to the diversity and wonder of the dinosaur era.

Chapter Six

Apatosaurus

Often confused with its close relative Brachiosaurus, Apatosaurus was another giant sauropod dinosaur that roamed the Earth during the Jurassic period.

The Apatosaurus, a long-necked giant, emerged from the shadows of scientific obscurity thanks to a lucky discovery in Colorado. In 1877, Arthur Lakes, a local miner with a keen eye for fossils, stumbled upon a treasure trove of bones in the Morrison Formation. These impressive remains belonged to a creature unlike anything seen before.

Lakes alerted Othniel Charles Marsh, a prominent paleontologist engaged in a heated rivalry with Edward Drinker Cope (the infamous "Bone Wars"). Recognizing the significance of the find, Marsh meticulously excavated the site, unearthing a partial skeleton. He named it Apatosaurus ajax, meaning "deceptive lizard" due to some initial confusion about the creature's anatomy.

This discovery marked a turning point in understanding sauropods, the long-necked giants of the Jurassic period. The Apatosaurus, with its immense size and herbivorous diet, provided valuable insights into the ecosystems of prehistoric North America.

Apatosaurus

Unique Features

Let's explore the unique features of this long-necked titan:

a. **Massive size:** Apatosaurus was a colossal creature, reaching lengths of up to 90 feet and weighing around 30 to 60 tons! While not as tall as Brachiosaurus, their sheer size was still awe-inspiring.

b. **Elongated neck and tail:** Apatosaurus boasted an incredibly long neck, though proportionally shorter than Brachiosaurus'. However, they made up for it with a tremendously long and whip-like tail, which could serve multiple purposes.

c. **Strong legs:** Like Brachiosaurus, Apatosaurus had powerful legs for supporting its enormous body weight.

d. **Bony ridges:** Apatosaurus had low bony ridges running along its back and tail, which may have provided muscle attachment points or served some regulatory function.

e. **Leaf-shaped teeth:** Their jaws were lined with numerous small, peg-like teeth perfect for tearing and stripping leaves from trees.

Diet

Apatosaurus, like its long-necked cousins, was a herbivore that feasted on a variety of plants. Their long necks allowed them to reach leaves from the tops of tall trees, such as conifers, cycads, and early flowering plants. To fuel their immense bodies, Apatosaurus needed to consume vast quantities of vegetation every day.

Other Facts about Apatosaurus

Here are some other interesting facts about Apatosaurus:

- The long, whip-like tail of Apatosaurus was used for defense against predators. This is because they could swing it with great force to deter other carnivorous dinosaur attackers like the Tyrannosaurus.
- Some evidence suggests Apatosaurus may have lived in herds, possibly for protection and easier access to food sources.
- Interestingly, Apatosaurus was originally named Brontosaurus, but due to a naming mix-up, the older name Apatosaurus took precedence.

In a nutshell, Apatosaurus was a magnificent herbivore that played a vital role in its ancient ecosystem. Their long necks and potentially social behavior continue to spark our curiosity about these gentle giants.

Chapter Seven

Pteranodon

While often mistaken for dinosaurs, Pteranodons were actually flying reptiles that lived alongside the giants of the prehistoric world.

The pterosaur, also known as Pteranodon wasn't the first to take flight in the fossil record, but its discovery in 1871 marked a significant leap in our understanding of these flying reptiles. The story unfolds in the dusty plains of Kansas, where paleontologist Othniel Charles Marsh, amidst the "Bone Wars" rivalry with Edward Drinker Cope, unearthed partial wing bones unlike anything documented before.

These bones, found in the Late Cretaceous Smoky Hill Chalk deposits, were unlike those of known birds or reptiles. Marsh, initially mistaken about the presence of teeth, recognized the bones' significance. His relentless pursuit for more fossils yielded a partial skull in 1876, revealing a toothless creature with a bony crest. This discovery, along with subsequent findings, solidified the Pteranodon's place as a giant pterosaur with a wingspan exceeding 7 meters, soaring through the Late Cretaceous skies.

Pterasaur

Physical Features

Let's explore the fascinating features of these aerial masters:

a. **Wingspan:** Pteranodons boasted an incredible wingspan, reaching up to an astounding 75 feet wide! This makes them one of the largest flying animals ever to have existed.

b. **Bony crest:** A hallmark feature of Pteranodons is the large bony crest extending from the top of their head. The function of this crest is still debated, but it may have served for display, steering, or heat regulation.

c. **Hooked beak:** Pteranodons lacked teeth and had a long, pointed beak. This beak was well-suited for snatching fish and other small animals from the air or water.

d. **Lightweight body:** Despite their large wingspan, Pteranodons had a relatively lightweight body structure with hollow bones, making flight easier and possible.

e. **Powerful flight muscles:** Attached to the bony crest, Pteranodons had powerful flight muscles that enabled them to soar through the air currents.

It is important to note that Pteranodons were not actually dinosaurs. They belonged to a different group of reptiles called pterosaurs.

Diet

Here's a glimpse into what Pteranodons likely ate:

a. **Fish:** Soaring over water, Pteranodons could swoop down to snatch fish from the surface.

b. **Small creatures:** They also fed on lizards, insects, and other small animals they could catch in their beaks.

Interesting facts about Pteranodons

Some other interesting facts about Pteranodons include:

- While Pteranodons could flap their wings to some extent. However, they were more adept at gliding on air currents.
- Having keen eyesight was crucial for Pteranodons to spot prey from the air while they fly.
- Pteranodons shared the skies with dinosaurs during the Cretaceous period, but they weren't closely related.

Note that Pteranodons were magnificent flying reptiles that ruled the skies during their time. Their incredible wingspan and adaptations for aerial life continue to amaze us today.

Chapter Eight

Spinosaurus

Spinosaurus has captured the imagination of paleontologists and dinosaur enthusiasts alike. This giant Theropod might have even rivaled the mighty T. Rex in size and was a unique predator with fascinating features:

The Spinosaurus story is a tale of discovery, loss, and rediscovery. In 1912, paleontologist Richard Markgraf unearthed a partial skeleton in the Egyptian Bahariya Oasis. The enormous spines on the creature's back vertebrae captivated German paleontologist Ernst Stromer von Reichenbach, who named it Spinosaurus aegyptiacus, meaning "spine lizard of Egypt."

Stromer's description which was based on the incomplete skeletons depicted Spinosaurus as a terrestrial predator. However, tragedy struck in

1944. Allied bombing during World War II destroyed the original Spinosaurus remains, leaving only Stromer's descriptions and sketches.

For decades, Spinosaurus remained an enigma. New discoveries and re-evaluations of existing fossils in the late 20th and early 21st centuries painted a different picture. The spines were likely a sail, and Spinosaurus may have been a semi-aquatic predator, adapted for hunting fish and swimming in rivers.

The Spinosaurus story continues to unfold, highlighting the fragmentary nature of the fossil record and the ongoing quest to understand these prehistoric giants.

Spinosaurus

Physical Features

a. ***Size:*** Estimates suggest Spinosaurus could reach up to 50 feet in length and weigh around 7 tons. While not as massive as some sauropods, it was still a formidable predator and possibly the largest carnivore ever discovered.

b. ***Spiny sail:*** The most distinct feature of Spinosaurus is the giant fin or sail on its back, formed by elongated neural spines. The exact purpose of this sail remains debated, but it may have been for display, temperature regulation, or even attracting mates.

c. ***Long and powerful jaws:*** Spinosaurus had elongated, crocodile-like jaws lined with sharp, conical teeth. These jaws were well-suited for grasping and puncturing prey.

d. ***Powerful legs:*** Despite its large size, Spinosaurus had relatively long and powerful legs, indicating it could move faster than other giant theropods.

e. ***Possible swimming adaptations:*** Recent discoveries suggest Spinosaurus may have had adaptations for a semi-aquatic lifestyle, such as a flattened tail fin and webbed feet.

Diet

The diet of Spinosaurus is still being debated due to limited fossil evidence. However, here are some theories:

- ***Fish eater:*** Spinosaurus' long jaws and possible aquatic adaptations suggest it might have primarily fed on fish and other water creatures.
- ***Opportunistic predator:*** It may have also hunted dinosaurs and other land animals when the opportunity arose.

Interesting Facts about Spinosaurus

Some other interesting facts about Spinosaurus include:

- Compared to other dinosaurs, Spinosaurus fossils were only discovered in the early 20th century.
- The exact shape and size of the sail on Spinosaurus' back are still being debated due to incomplete fossils.
- The idea of Spinosaurus being a semi-aquatic predator is a relatively new discovery that challenges our understanding of this giant dinosaur.

Yet, Spinosaurus was a unique and fascinating predator that continues to be studied and debated. Its large size, unusual features, and possible lifestyle make it a truly remarkable creature from the prehistoric world.

Chapter Nine

Allosaurus

Reigning supreme during the Jurassic period, Allosaurus was a fearsome theropod dinosaur and a top predator in its ecosystem.

The Allosaurus, a fearsome predator with a name meaning "different lizard," emerged from the rocks of Colorado in 1877. Othniel Charles Marsh, a paleontologist entrenched in the famous "Bone Wars" rivalry with Edward Drinker Cope, unearthed vertebrae unlike any seen before. These bones had unique hollow spaces, different from those of known dinosaurs.

Intrigued by this discovery, Marsh embarked on a relentless excavation, unearthing more Allosaurus remains in Wyoming. The complete picture of the Allosaurus began to form – a massive predator with a powerful jaw and a long, muscular tail. This discovery, along with the mounting evidence from the Bone Wars, painted a thrilling picture of the Late Jurassic period in North America, dominated by these fearsome theropods.

Allosaurus

Physical Features

Let's delve into the features that made them such effective hunters:

a. **Sharp claws and teeth:** Allosaurus possessed powerful jaws lined with sharp, serrated teeth perfect for tearing flesh. Additionally, their distinctive curved claws on each foot were likely used for grasping and holding onto prey during a kill.

b. ***Powerful legs and strong build:*** Allosaurus had powerful hind legs, enabling them to run swiftly and chase down their prey. Their muscular build further supported their predatory lifestyle.

c. ***Lightweight skull:*** Compared to some theropods, Allosaurus had a lighter skull, which may have contributed to their agility and speed.

d. ***Binocular vision:*** Similar to T. Rex and other predatory dinosaurs, Allosaurus had eyes positioned forward on their heads, granting them depth perception crucial for hunting.

e. ***Size:*** While not as massive as giants like Brachiosaurus, Allosaurus were still formidable predators, reaching up to 30 feet in length and weighing around 2 tons.

Diet

Allosaurus were carnivores, and their diet likely consisted of:

a. ***Herbivores:*** They preyed on various herbivores that shared their habitat, including early sauropods, stegosaurs, and ornithischians.

b. ***Scavenging:*** Allosaurus may have also scavenged carcasses of other dinosaurs when the opportunity arose.

Interesting Facts about Allosaurus

Some other interesting facts about Allosaurus include:

- Allosaurus walked upright on two powerful legs, a common trait shared by most theropod dinosaurs.
- While solitary behavior is possible, some evidence suggests Allosaurus may have occasionally hunted in packs, making them even more effective predators.
- Allosaurus fossils were among the first dinosaur discoveries, providing valuable insights into theropod anatomy and evolution.

As a result of the above, Allosaurus played a crucial role in the Jurassic food chain, and their fossils continue to shed light on the fascinating world of prehistoric predators.

Chapter Ten

Diplodocus

Diplodocus, meaning "double-beamed lizard," was a giant sauropod dinosaur known for its immense size and staggering length.

The Diplodocus story begins not with a dramatic excavation, but with a single bone. In 1877, miners in Colorado unearthed a peculiar tail vertebra unlike anything documented. It found its way to paleontologist Benjamin Mudge, who recognized its odd features – a double beam structure unlike any dinosaur vertebra known at the time.

Intrigued, Mudge enlisted the help of paleontologist Othniel Charles Marsh, a prominent figure in the "Bone Wars" rivalry with Edward Drinker Cope. Marsh, recognizing the potential significance, named the creature Diplodocus longus, meaning "double-beamed long one" based on the tail vertebra.

Over the next few years, further discoveries in Colorado, Utah, and Wyoming brought more Diplodocus bones to light. The massive skeleton revealed a truly gigantic creature – one of the longest dinosaurs ever discovered. This long-necked, long-tailed herbivore, with its whip-like tail, captured the imagination of the scientific community and solidified our understanding of the diverse giants that roamed prehistoric North America.

Physical Features

Let's explore what made this gentle giant a marvel of the Jurassic period:

- a. **Sheer size:** Diplodocus was a colossal creature, reaching lengths of up to 90 feet and weighing around 20 to 30 tons! Their long, slender bodies were truly impressive.

b. ***Extraordinary length:*** The most striking feature of Diplodocus is the combination of its incredibly long neck and tail. The neck could reach up to 30 feet, allowing them to browse leaves from high trees inaccessible to most other dinosaurs. Their tails were even longer, stretching up to 50 feet and possibly used for balancing and defense.
c. ***Whip-like tail:*** The long tail of Diplodocus was muscular and flexible. While some theories suggest it may have been used as a weapon against predators, recent studies indicated it may have functioned primarily for balance and counterbalancing the weight of the neck.
d. ***Lightweight build:*** Despite their massive size, Diplodocus had a surprisingly lightweight skeleton due to hollow bones. This adaptation helped support their enormous bodies.
e. ***Small head:*** Diplodocus had a relatively small head compared to its body size. Their

jaws were lined with peg-shaped teeth suitable for tearing and stripping leaves.

Diet

The Diplodocus was herbivores, and their long necks were perfectly designed for their plant-based diet:

- **High browsers:** Their long necks allowed them to reach leaves from the canopy of tall trees, including ferns, cycads, and early flowering plants.
- **Passive feeders:** Due to their lack of sharp claws or teeth and their gentle nature, Diplodocus most likely spent their days peacefully grazing on leaves.

Interesting Facts about Diplodocus

Some other interesting facts about Diplodocus include:

- The Diplodocus was an herbivore and lacked the physical features for actively hunting other dinosaurs.
- Scientists believe Diplodocus may have lived in herds for protection and easier access to food sources.
- The name "Diplodocus" refers to the distinctive double beams formed by the lower bones running along the underside of their tail.

Thus, the Diplodocus, were magnificent herbivores that played a vital role in their ancient ecosystem. Their immense size, long necks, and peaceful nature continue to spark our curiosity about these gentle giants.

Chapter Eleven

Parasaurolophus

Parasaurolophus stands out among dinosaurs with its unique bony crest adorning its head.

The Parasaurolophus, a bizarre crested herbivore with a name meaning "near crested lizard," wasn't always recognized for its extraordinary headgear. The story begins in 1920, amidst the Canadian badlands of Alberta. A field crew from the University of Toronto unearthed a partial skeleton, including a skull unlike anything seen before.

This skull possessed a long, hollow crest curving up and back from the head. Paleontologist William Parks, initially unsure of its function, named the creature Parasaurolophus based on its resemblance to Saurolophus, a known dinosaur with a crest.

The crest of the Parasaurolophus sparked scientific curiosity. Theories ranged from sound amplification for communication to a display for attracting mates or intimidating rivals. The discovery of Parasaurolophus remains in Alberta, Utah, and New Mexico over the years provided more clues, but the true purpose of the crest remains a captivating mystery, fueling our fascination with this unique dinosaur.

Parasaurolophus

Physical Features

This fascinating herbivore had several interesting features and played a vital role in its prehistoric environment and these include:

a. **Distinctive crest:** The most striking feature of Parasaurolophus is the large, hollow crest on its head. This crest could vary in shape and size depending on the species, but it was a defining characteristic.

b. **Hollow bones:** Like many other sauropods, Parasaurolophus had hollow bones to help support its massive body weight.

c. **Beak and teeth:** Their jaws were equipped with a beak at the front, ideal for nipping leaves, and rows of flat teeth suitable for grinding plant material.

d. **Size:** Parasaurolophus wasn't as gigantic as some sauropods, reaching lengths of up to 40 feet and weighing around 2.5 tons.

e. **_Very Agile:_** Despite its size, Parasaurolophus was a relatively agile browser due to its long legs and flexible neck.

Diet

Some of what Parasaurolophus likely ate includes:

- **_Low-lying plants:_** Their diet primarily consisted of ferns, cycads, and other low-growing vegetation.
- **_Selective feeders:_** Their beaks allowed them to be more selective feeders, choosing leaves of specific plants.

Note that the most intriguing aspect of Parasaurolophus is its head crest. The leading theory suggests the hollow crest functioned as a resonator, amplifying sounds produced by the dinosaur. This would have allowed them to communicate with each other over long distances, possibly for finding mates or warning others of danger. The crest may have also served other purposes, such as display for attracting mates or regulating body temperature.

Facts about Parasaurolophus

Some interesting facts about Parasaurolophus include:

- Scientists believe Parasaurolophus may have lived in herds for protection and easier access to food sources.
- Different Parasaurolophus species sported crests of varying shapes and sizes, hinting at potential differences in communication or social structures.
- The first Parasaurolophus fossils were only discovered in the early 20th century.

As a whole, Parasaurolophus was a unique herbivore with a fascinating adaptation which is its head crest. The possible use of this crest for communication adds another layer of complexity and intrigue to the world of dinosaurs.

Chapter Twelve

Ankylosaurus

Imagine a living tank covered in bony plates and spikes – that's the Ankylosaurus! This heavily armored herbivore roamed the Earth during the Late Cretaceous period and was a marvel of defensive adaptations.

The Ankylosaurus, a heavily armored tank of the dinosaur world, emerged from the dusty plains of Montana in 1906. Paleontologist Barnum Brown, the same explorer who unearthed the fearsome Tyrannosaurus Rex, led an expedition that unearthed the first Ankylosaurus remains in the Hell Creek Formation.

These initial finds were far from complete – a partial skull, scattered armor plates, and vertebrae. Brown, ever the determined explorer, continued his quest. Four years later, he unearthed more

Ankylosaurus bones in Alberta, Canada, including the first known tail club – a bony weapon that would become synonymous with this armored herbivore.

Despite these discoveries, the Ankylosaurus remains somewhat enigmatic. Unlike some dinosaurs with numerous well-preserved skeletons, Ankylosaurus fossils are scarce and incomplete. This scarcity has paleontologists piecing together the puzzle of its appearance, with the exact arrangement of its armor plates still debated. However, the available evidence paints a clear picture – the Ankylosaurus was a heavily armored dinosaur, well-equipped to deter the predators of its time.

Ankylosaurus

Physical Features

a. ***Spiked armor:*** The defining feature of Ankylosaurus is its incredible armor. Their bodies were covered in thick, bony plates called osteoderms, forming a nearly impenetrable shield.

b. ***Spiked tail club:*** Ankylosaurus possessed a powerful tail tipped with a massive bony club. This club could be swung with great force and served as a formidable weapon against any attacker.

c. ***Low profile:*** Ankylosaurus had a short, stocky body with short legs, keeping its center of gravity low and making it difficult for predators to topple over.

d. ***Limited neck movement:*** Due to the armor on their necks, Ankylosaurus tends to have restricted movement in that area.

e. ***Small head:*** Their heads were relatively small compared to their bodies, with a beak-like mouth perfect for tearing and chewing tough plants.

Diet

While heavily armored, Ankylosaurus were plant-eaters feasting on low-lying vegetation. Their diet consisted of ferns, cycads, and other low-growing plants accessible with their beaks.

Despite their small heads, Ankylosaurus had powerful jaws and grinding teeth, allowing them to break down tough plant material.

Other Facts about Ankylosaurus

Some interesting facts about Ankylosaurus include:
- Some theories suggest the bony plates on their backs may have helped regulate body temperature.
- Ankylosaurus likely lived solitary lives, relying on their armor for protection.

- Recent fossil discoveries suggest Ankylosaurus may have inhabited islands, explaining their unique evolutionary adaptations.

Ankylosaurus were fascinating herbivores that carved out a successful niche in their environment. Their remarkable armor and defensive features continue to amaze us today, earning them the nickname "armored tank" of the dinosaur world.

Chapter Thirteen

Iguanodon

Iguanodon holds a special place in paleontological history, being one of the first dinosaur discoveries to capture the public imagination. However, our understanding of this early herbivore has evolved significantly since its initial identification.

The story of Iguanodon began in the early 1820s when Mary Ann Mantell, an amateur geologist, discovered teeth unlike any seen before. These discoveries, further studied by her husband Gideon Mantell, laid the foundation for naming Iguanodon in 1825, making it one of the first formally described dinosaurs. However, due to limited fossil evidence at the time, early reconstructions of Iguanodon were inaccurate. For instance, they were mistakenly depicted as bipedal creatures resembling giant iguanas, with the thumb spike misidentified as a horn on the snout.

However, as more fossils came to light, the picture of Iguanodon became clearer. Scientists discovered they were actually quadrupedal herbivores with the characteristic thumb spike on their forelimbs, possibly used for defense or foraging.

Iguanodon

Physical Features

A closer look at the physical features of Iguanodon will spot:

a. **Large herbivores:** Iguanodon grew to be around 30 feet long and weighed up to several tons, making them formidable herbivores during their time.

b. **Powerful jaws and teeth:** Their jaws were lined with specialized teeth well-suited for grinding tough plant material.

c. **Thumb spike:** The most unique feature of Iguanodon is the large, bony spike protruding from their thumbs. While the exact function remains debated, it may have been used for defense against predators or for browsing vegetation.

d. **Potential for bipedalism:** Recent studies suggest Iguanodon, while primarily quadrupedal, might have been able to stand on their hind legs for short periods, possibly for reaching high foliage.

Diet

Iguanodon's diet consisted mainly of plant material:
a. **Leaves:** Their teeth and jaw structure indicate a diet of leaves from cycads, ferns, and potentially early flowering plants.
b. **Grazing:** Iguanodon likely spent a significant portion of their day foraging and consuming large quantities of vegetation.

Further Facts about Iguanodon

Some interesting facts about Iguanodon include:
- Over time, paleontologists have identified several Iguanodon species based on fossil variations, showcasing the diversity within this group.

- While solitary behavior is possible, some evidence suggests Iguanodon may have lived in herds for protection and easier access to food sources.
- The legacy of Iguanodon is intertwined with ongoing scientific debate. A case in point is in the classification of some fossils previously attributed to Iguanodon which had now been revised as paleontologists gain a deeper understanding of this dinosaur group.

Thus Iguanodon's early discovery and the misconceptions surrounding it highlight the ongoing process of scientific research and refinement. This fascinating herbivore continues to be an important part of our understanding of dinosaur evolution and diversity.

Chapter Fourteen

Megalosaurus

Megalosaurus holds a unique position in dinosaur history. For instance, it was one of the very first dinosaurs to be scientifically described, paving the way for further discoveries and shaping our understanding of these prehistoric giants. However, much about Megalosaurus remains shrouded in mystery due to limited and fragmentary fossil evidence.

The story of Megalosaurus began in the early 19th century when William Buckland, a renowned geologist, unearthed large leg bones near Oxford, England. Recognizing their distinct characteristics, Buckland named the creature Megalosaurus in 1824, meaning "great lizard" – one of the first formal designations of a dinosaur.

Unfortunately, the initial Megalosaurus discovery consisted primarily of leg bones and fragments of jaw and teeth. This limited evidence made it challenging to fully reconstruct the anatomy and lifestyle of this dinosaur.

Based on the available fossils, Megalosaurus was initially depicted as a giant, bipedal predator resembling a large lizard. However, as more dinosaur discoveries were made, this image needed revision.

Megalosaurus

Physical Features

A glimpse into what we know about Megalosaurus based on current understanding:

a. ***Large theropod:*** Fossil evidence suggests Megalosaurus belonged to the theropod group, a category encompassing bipedal carnivorous dinosaurs. Size estimates place them at around 20 to 30 feet long and several tons in weight.

b. ***Powerful jaws and teeth:*** The available jaw fragments indicate Megalosaurus possessed sharp, serrated teeth well-suited for tearing flesh.

c. ***Possible scavenger:*** Due to the limited knowledge about their forelimbs and the presence of other larger theropods during their time, some paleontologists theorize Megalosaurus might have been scavengers, feeding on the carcasses of other dinosaurs.

d. ***Potential for hunting:*** While scavenging is a possibility, Megalosaurus likely also actively hunted smaller prey based on their theropod classification and powerful jaws.

Diet

Due to the fragmentary nature of the fossils, pinpointing the exact diet of Megalosaurus is challenging. However, they are believed to have carnivorous tendencies as their teeth structure strongly suggests a meat-eating diet. Also, depending on their size and agility, Megalosaurus may have preyed on smaller dinosaurs, reptiles, and early mammals.

Further Facts about Megalosaurus

Some interesting facts about Megalosaurus include:
- Over the years, as new dinosaur discoveries were made, the classification of Megalosaurus

has been revised. Fossils previously attributed to Megalosaurus might belong to different theropod groups.

- Despite ongoing research, the lack of a complete Megalosaurus skeleton makes it difficult to definitively determine their physical features and behavior.
- Megalosaurus, though shrouded in mystery, played a crucial role in sparking scientific interest in dinosaurs. For example, its early description laid the foundation for further exploration and understanding of these magnificent creatures.

In a nutshell, Megalosaurus serves as a reminder of the ongoing process of paleontological discovery. While limited fossil evidence leaves many questions unanswered, this dinosaur holds historical significance for igniting our fascination with the prehistoric world.

Chapter Fifteen

Camarasaurus

Camarasaurus, meaning "chambered lizard," was a sauropod dinosaur that roamed the Earth during the Late Jurassic period. They were known for their immense size and incredibly long necks, making them gentle giants of the prehistoric world.

The story of Camarasaurus unfolds in Colorado, amidst the dusty landscapes of the Morrison Formation. The year was 1877, a period marked by intense fossil discoveries fueled by the "Bone Wars" rivalry between paleontologists Othniel Charles Marsh and Edward Drinker Cope. In this climate of competition, Oramel W. Lucas, working for Marsh, stumbled upon a treasure trove of bones – vertebrae and fragments unlike anything documented before.

These initial scraps belonged to a gigantic creature, hinting at a sauropod dinosaur – a long-necked herbivore. The frenzy to unearth more intensified. Marsh, recognizing the significance of the find, named the creature Camarasaurus supremus, meaning "superior chambered lizard" based on the hollow spaces in its vertebrae, a weight-saving adaptation.

Over the years, more Camarasaurus remains came to light in Colorado, Utah, Wyoming, and New Mexico. These discoveries revealed a long-necked, long-tailed dinosaur with a distinctive, high-set head. Unlike its gigantic cousins like Diplodocus, Camarasaurus was smaller and browsing on higher foliage. This unique feature, along with its abundant fossils, has made Camarasaurus one of the most well-understood sauropods, providing valuable insights into the diverse herbivores that roamed prehistoric North America.

Physical Features

a. ***Massive size:*** Camarasaurus wasn't the largest sauropod, but they were still impressive creatures. For instance, they could grow up to 85 feet long and weigh around 20 tons, requiring a vast amount of plant material to sustain itself.

b. ***Elongated neck:*** One of the most defining features of Camarasaurus is their incredibly long neck, making up almost half their body length! This adaptation allowed them to reach leaves from the tops of tall trees, inaccessible to most other dinosaurs.

c. ***Small head:*** Despite their large size, Camarasaurus had a relatively small head with a blunt snout. Their jaws were lined with numerous small, peg-like teeth perfect for tearing and stripping leaves.

d. ***Powerful legs:*** Supporting their enormous bodies were thick, pillar-like legs. While their front legs were slightly shorter than their back legs, they were still strong enough to hold their weight and allow for movement.

e. **Bony plates:** Some Camarasaurus species had bony plates embedded in their skin along their neck, back, and tail. The function of these plates is still debated, but they may have provided some level of protection or aided in heat regulation.

Diet

Camarasaurus were herbivores, and their long necks were perfectly suited for their plant-based diet. For instance, their long necks allowed them to reach foliage from the highest branches of trees, including conifers, cycads, and early flowering plants. Also, the small size and shape of their teeth suggest Camarasaurus were selective feeders, choosing leaves of specific plants.

Other Facts about Camarasaurus

Here are some other interesting facts about Camarasaurus:

- Due to their plant-eating diet and lack of sharp claws or teeth, Camarasaurus were most likely gentle giants.
- Some scientists believe Camarasaurus may have lived in herds for protection and easier access to food sources.
- Their large bodies likely housed complex digestive systems to efficiently process the vast amounts of plant material they consumed.
- Camarasaurus fossils have been found in North America, Europe, and even Africa, suggesting they were widespread during the Late Jurassic period.

Camarasaurus were fascinating sauropods that played a vital role in their ecosystem. Their long necks and adaptations for browsing high foliage helped them thrive in a world dominated by giant plants. While not the largest dinosaurs, their size and unique features continue to capture our imagination.

Chapter Sixteen

Corythosaurus

Corythosaurus, meaning "helmet lizard," was a fascinating duck-billed dinosaur (hadrosaur) that lived during the Late Cretaceous period. Their most striking feature was a large, hollow crest adorning their heads, making them stand out among other herbivores.

As already stated, the most intriguing aspect of Corythosaurus is their head crest. The leading theory suggests the hollow crest functioned as a resonator, amplifying sounds produced by the dinosaur. This would have allowed them to communicate with each other over long distances, possibly for finding mates or warning others of danger. The crest may have also served other purposes as well, such as display for attracting mates or even heat regulation.

Corythosaurus

Features

- ***Distinctive crest:*** The most remarkable characteristic of Corythosaurus is the prominent bony crest on their heads. This crest was hollow and extended upwards and backwards, resembling a helmet or a Corinthian helmet worn by ancient Greek the warriors.
- ***Duck-billed jaws:*** Like other hadrosaurs, Corythosaurus had a broad, flat beak at the front of their jaws. Their jaws were lined with numerous small, peg-like teeth well-suited for grinding tough plant material.
- ***Hollow bones:*** A common feature among many large dinosaurs, Corythosaurus had hollow bones to help reduce their body weight while maintaining immense size.
- ***Size:*** While not the largest herbivores, Corythosaurus were still impressive creatures,

reaching lengths of up to 33 feet and weighing around 5 tons.
- **Powerful legs:** Their legs were strong and muscular, allowing them to support their massive bodies and move efficiently.

Diet

Corythosaurus were herbivores, and their unique features played a crucial role in their plant-based diet. Their broad snouts and low grazing posture suggest they fed on low-lying plants like ferns, cycads, and flowering shrubs. Also, the numerous small teeth in their jaws were perfect for grinding down tough plant material.

Other Interesting Facts about Corythosaurus

- Some evidence suggests Corythosaurus may have lived in herds for protection and easier access to food sources.

- Different Corythosaurus species sported crests of varying shapes and sizes, hinting at potential differences in communication or social structures.
- The first complete Corythosaurus skull wasn't discovered until the early 20th century, providing valuable insights into their head anatomy and the potential functions of the crest.

Corythosaurus were remarkable herbivores with a unique adaptation – their head crest. The possibility of this crest being used for communication adds another layer of complexity and intrigue to the world of dinosaurs. Their duck-billed jaws and plant-eating lifestyle solidify their role as successful herbivores in their ancient ecosystem.

Chapter Seventeen

Carnotaurus

Carnotaurus, meaning "meat-eating bull," was a truly unique theropod dinosaur that roamed South America during the Late Cretaceous period. Their most striking features were the horns above their eyes and the unusual bumps covering their bodies, making them a sight to behold.

The Carnotaurus story is unique because it features not just a single discovery, but a single dinosaur. In 1984, paleontologist José Bonaparte unearthed a nearly complete skeleton in Argentina's Chubut Province. This newfound theropod, christened Carnotaurus sastrei (meaning "meat-eating bull of Sastre"), was unlike any seen before.

The most striking feature was a pair of short, horns curving above its eyes. Theories abound about their function – for combat with rivals, for attracting mates, or perhaps for display. The discovery also included fossilized skin impressions, a rare find that provided valuable clues about the Carnotaurus' appearance.

This single, remarkably well-preserved Carnotaurus skeleton has become the cornerstone of our understanding of this peculiar theropod. It sheds light on the diversity of meat-eaters that roamed South America during the Late Cretaceous period, showcasing the fascinating adaptations that dinosaurs developed across the globe.

Carnotaurus

Physical Features

a. ***Distinctive horns:*** The most recognizable feature of Carnotaurus is the presence of a pair of large, bony horns protruding from above their eyes. The exact function of these horns remains debated, but they may have been used for display during mating rituals or for fighting other predators.

b. ***Bumpy body:*** Another distinctive characteristic of Carnotaurus is their bumpy or pebbled skin texture. Fossil evidence suggests their bodies were covered in small, bony nodules, giving them a rough and uneven appearance.

c. ***Powerful jaws and teeth:*** Carnotaurus possessed a deep, narrow skull equipped with sharp, serrated teeth. These adaptations indicate they were formidable predators.

d. ***Small arms:*** One of the most intriguing features of Carnotaurus is their incredibly small forelimbs, even compared to other theropods. The arms were barely functional, raising questions about their purpose in hunting.

e. ***Strong legs and powerful bite:*** Carnotaurus compensated for their small arms with powerful legs and a strong bite. Their legs allowed them to run swiftly, while their jaws could deliver a crushing force.

Diet

Here's a look at what Carnotaurus likely ate:

- Their teeth and jaw structure clearly indicate Carnotaurus were carnivores.
- They likely preyed on various herbivores that shared their habitat, including sauropods, ornithischians, and possibly smaller theropods.
- While the exact details are unclear due to the limitations of fossil evidence, some theories

suggest Carnotaurus might have relied on speed and their powerful bite to take down prey. Their small arms may have been used for grasping or holding onto struggling prey.

Other Interesting Facts about Carnotaurus

- Despite their bulky bodies, Carnotaurus were likely capable of running quite fast due to their powerful legs.
- Scientists believe Carnotaurus were solitary predators, hunting and feeding alone.
- Paleontologists are fortunate to have a nearly complete Carnotaurus skeleton, providing valuable insights into their anatomy and lifestyle.

In a nutshell, Carnotaurus stands out as a unique predator with an unusual combination of features. Their horns, bumpy skin, and small arms continue to spark curiosity and debate among paleontologists. While the exact function of some of these traits remains a mystery, Carnotaurus serves as a fascinating example of dinosaur diversity and adaptation during the prehistoric era.

Chapter Eighteen

Majungasaurus

Imagine a fearsome theropod dinosaur with a unique bony crest on its head – that's Majungasaurus! This predator, known as the "giant lizard from Mahajanga," ruled the lands of Madagascar during the Late Cretaceous period.

The story of Majungasaurus is a tale of fragmentary finds and hidden clues. The first hint emerged in 1895, not from a complete skeleton, but from scattered bones along the Betsiboka River in Madagascar. French paleontologist Charles Depéret unearthed vertebrae, a claw, and teeth, enough to tantalize but not enough to identify a clear picture.

These remains were initially assigned to Megalosaurus, a wastebasket taxon for various large theropods at the time. The story took a backseat for decades, with only fragmentary finds adding to the puzzle. It wasn't until 1955 that René Lavocat, another paleontologist, unearthed a crucial piece – a lower jawbone with distinctive teeth. This jaw, along with the earlier finds, led Lavocat to create a new genus: Majungasaurus, named after the region where it was found.

Even with the new genus, the true nature of Majungasaurus remained elusive for decades. It wasn't until the late 20th century that more complete skeletons came to light, revealing a fearsome predator with a short snout and a bony crest on its head. Interestingly, some fossils even showed bite marks on other Majungasaurus bones, hinting at possible cannibalistic behavior – a grisly detail that solidified its reputation as a formidable predator.

The story of Majungasaurus continues to unfold as more fossils are unearthed. But even with the piecemeal discoveries, this dinosaur offers a window into the unique fauna of Madagascar and the fascinating diversity of theropods that existed during the Late Cretaceous period.

Majungasaurus

Physical Features

a. **Distinctive crest:** The most striking feature of Majungasaurus is the large, bony crest adorning their skull. This crest was likely a display feature, possibly used for attracting mates or asserting dominance within their species.

b. **Powerful build:** Majungasaurus were formidable predators with robust bodies built for taking down prey. Their large skulls housed powerful jaws and sharp, serrated teeth.

c. **Short arms:** Similar to Carnotaurus, the forelimbs of Majungasaurus were surprisingly small compared to their hind legs. The function of these short arms remains debated, but they may have played a role in grasping or manipulating prey during a kill.

d. **Strong legs and sharp claws:** Their powerful legs enabled them to run swiftly, while the sharp claws on their feet were likely used for tearing flesh.

e. **Size:** Majungasaurus were not the largest theropods, reaching lengths of up to 30 feet and weighing around 2 tons. However, their size and adaptations made them effective predators in their environment.

Diet

Their teeth and jaw structure clearly indicate they were meat-eaters. They likely preyed on various herbivores that shared their habitat, including dinosaurs like sauropods and ornithischians. Also, some evidence suggests Majungasaurus might have also scavenged carcasses of other dinosaurs, supplementing their diet with readily available food sources.

Other Interesting Facts about Majungasaurus

- Fossils of Majungasaurus have only been found in Madagascar, suggesting they were island dwellers. This isolation may have contributed to the development of unique anatomical features.
- Scientists believe Majungasaurus were likely solitary animals, except possibly during mating season.
- Compared to other dinosaurs, Majungasaurus fossils were only discovered in the late 20th century. Further research is ongoing to gain a deeper understanding of this fascinating predator.

Thus Majungasaurus represents a unique theropod dinosaur with an intriguing combination of traits. Their distinctive head crest, powerful build, and short arms make them a subject of continued study. This crested predator adds to the diversity of carnivores that thrived during the Late Cretaceous period.

Chapter Nineteen

Baryonyx

Baryonyx, meaning "heavy claw," was a unique theropod dinosaur that carved a niche as a specialized fish-eater during the Early Cretaceous period. Their most distinctive feature, a massive hooked claw on their thumb, sets them apart from other known theropods.

The Baryonyx story is an intriguing tale of a dinosaur initially mistaken for a giant claw. In 1951, English paleontologist William Walker found a massive, curved claw unlike anything documented before in the Surrey woodlands of England. Believing it belonged to a giant marine reptile, he named it Baryonyx nghĩa là "heavy claw."

For decades, Baryonyx remained an enigma. It wasn't until 1986 that paleontologist Alan Charig re-examined the claw alongside other scattered bones found in the same location. This meticulous analysis revealed a surprising truth – the bones belonged together, forming a new kind of theropod dinosaur.

The rediscovery of Baryonyx was a turning point. The massive claw, now recognized as a unique adaptation for hunting, sparked scientific curiosity. Further excavations in the following years unearthed more Baryonyx remains, including a partial skull with a distinctive, downward-curving snout.

This bizarre feature, along with sharp, recurved teeth, suggested Baryonyx may have been a fish-eater – a departure from the typical meat-eating theropods. The discovery of Baryonyx challenged existing ideas about dinosaur diets and behavior, highlighting the diversity of adaptations that thrived during the Early Cretaceous period.

Baryonyx

Other Physical Features

a. **Hooked thumb claw:** The defining characteristic of Baryonyx is the enormous, curved claw on their single thumb. This unique adaptation, unlike the sharp claws seen in most theropods, it was likely used for snagging and holding slippery prey.
b. **Elongated snout:** Baryonyx possessed a long and narrow snout, somewhat resembling that of a crocodile. This snout was lined with small, sharp teeth ideal for catching and gripping fish.
c. **Powerful jaws:** Their jaws were strong and well-suited for crushing the bones of their prey.
d. **Strong legs and flexible body:** While not the largest theropods, Baryonyx had powerful legs for moving swiftly and a flexible body that may have aided in maneuvering near water.

e. **Size:** Baryonyx grew to be around 30 feet in length and weighed up to 2 tons, making them formidable predators in their environment.

Major Diet

Unlike most theropods known for hunting terrestrial animals, Baryonyx had a specialized diet. The physical adaptations of Baryonyx strongly suggest they were primarily fish-eaters. Their hooked claw, elongated snout with sharp teeth, and possible adaptations for wading in water make this evident. The hooked claw was likely used to snag fish from the water, while their sharp teeth and strong jaws allowed them to crush bones and consume their prey effectively.

Other Interesting Facts about Baryonyx

- While Baryonyx could walk on land, their anatomy suggests they likely spent a

significant amount of time near water bodies, hunting and feeding on fish.
- Scientists believe Baryonyx were solitary predators, hunting and feeding alone.
- The first Baryonyx fossils were unearthed in England in the 1980s, providing paleontologists with a glimpse into this unique fish-eating dinosaur.

Baryonyx stands out as a remarkable example of adaptation in the dinosaur world. Their hooked claw, elongated snout, and possible semiaquatic lifestyle demonstrate how dinosaurs evolved to thrive in specific ecological niches. Their discovery continues to challenge our understanding of the diverse feeding strategies employed by these prehistoric creatures.

Chapter Twenty

Giganotosaurus

Giganotosaurus, meaning "giant southern lizard," lived up to its name. This colossal theropod dinosaur ruled the Late Cretaceous period in what is now South America and was a sight to behold.

Giganotosaurus, the "giant southern lizard," wasn't unearthed in a single dramatic excavation. Its story began with a single bone hinting at an immense creature. In 1993, Rubén Dario Carolini, an amateur fossil hunter in Argentina's Patagonia region, spotted a protruding tibia unlike any he'd seen before. This chance discovery sparked a paleontological odyssey.

Carolini alerted paleontologists at the National University of Comahue. News of the find reached Rodolfo Coria and Leonardo Salgado, who led an excavation that yielded a treasure trove – a massive, fragmented skeleton. It belonged to a theropod dinosaur unlike any other, exceeding even the fearsome Tyrannosaurus Rex in size.

The unearthed bones, though scattered, were enough to send shockwaves through the paleontological world. Coria and Salgado named the creature Giganotosaurus carolinii in 1995, honoring both the immense size of the animal and the man who first unearthed it.

Giganotosaurus' discovery challenged our understanding of theropod size and diversity. This giant predator, with its powerful jaws and sharp teeth, reigned supreme in the Late Cretaceous ecosystems of South America, captivating our imagination with its glimpse into the power and ferocity of the prehistoric world.

Giganotosaurus

Physical Features

a. Massive Size: Giganotosaurus was one of the largest known land predators ever to have existed. Size estimates suggest they could reach lengths of up to 43 feet and possibly weigh around 13 tons, rivaling even the mighty Tyrannosaurus Rex.

b. **Powerful build:** Their bodies were robustly built, with massive skulls housing powerful jaws and sharp, serrated teeth. Their legs were strong and well-suited for supporting their immense weight.

c. **Unique adaptations:** Despite their large size, Giganotosaurus had some adaptations for agility, such as a relatively small head compared to their body size and possibly three-toed feet.

d. ***Predatory lifestyle:*** There's no doubt that Giganotosaurus was a ferocious carnivore at the top of the food chain. Their physical features were perfectly designed for taking down large prey.

Diet

They likely preyed on various herbivores that shared their habitat, including sauropods, ornithischians, and possibly even other theropods. The details of their hunting behavior are under debate, but their powerful jaws and sharp teeth suggest they could inflict devastating bites. Their size and strength would have been a significant advantage when overpowering prey.

Other Intriguing Aspects about Giganotosaurus

- Fossil evidence suggests some Giganotosaurus individuals might have possessed deeper

snouts than others, hinting at potential variations within the species.
- Giganotosaurus fossils have primarily been found in South America, providing valuable insights into the unique fauna of that continent during the Late Cretaceous.
- While much has been learned about Giganotosaurus, further studies continue to shed light on their behavior, hunting strategies, and the ecological role they played in their environment.

Giganotosaurus and T. Rex Comparison

Giganotosaurus and Tyrannosaurus Rex are often compared due to their immense size and dominance as apex predators in the following areas:

i. **Length:** Giganotosaurus: Up to 43 feet; T. Rex: Up to 40 feet.

ii. **Weight:** Giganotosaurus: Up to 13 tons (estimates vary); T. Rex: Up to 8 tons (estimates vary).

Note that while Giganotosaurus might have had a slight edge in size, both were formidable predators. Also, factors like bite force, hunting strategies, and the specific prey they targeted would also play a role in determining their effectiveness as predators.

Giganotosaurus stands as a testament to the incredible diversity and scale of theropod dinosaurs. Their immense size, powerful build, and adaptations for hunting solidify their position as top predators of their time. While questions remain about their specific behavior and ecological role, Giganotosaurus continues to spark our imagination and serves as a reminder of the fascinating creatures that once roamed the Earth.

Conclusion

As we reach the end of our exploration, we stand at the precipice of a bygone era – the world of dinosaurs. These magnificent creatures, giants of a bygone time, continue to capture our imagination with their diversity, adaptations, and sheer scale.

This book has unveiled a mere glimpse into the incredible story of dinosaurs. We have encountered herbivores with remarkable adaptations for browsing high foliage, massive predators with bone-crushing jaws, and even specialized fish-eating theropods. Each species showcased the remarkable ability of life to adapt and thrive in a dynamic environment. However, the extinction of dinosaurs millions of years ago marks a significant turning point in Earth's history. Yet, their legacy lives on in the form of fossilized remains that continue to be unearthed and studied.

Also, our understanding of dinosaurs is constantly evolving as new discoveries are made and scientific research progresses. Each fossil fragment, each footprint, each tooth tells a story, offering a window into the lives of these ancient creatures.

Remember, the journey of dinosaur exploration is far from over and as we delve deeper into the past, we unlock new knowledge about their behavior, social interactions, and the ecosystems they inhabited. This pursuit not only expands our understanding of the prehistoric world but also sheds light on the broader story of life on Earth.

Understand that dinosaurs may be extinct, but their enduring appeal serves as a reminder of the incredible diversity of life that has graced our planet. Their story is a testament to the power of evolution, the delicate balance of ecosystems, and the awe-inspiring wonders that nature can produce.

As we move forward, the lessons learned from studying dinosaurs can inform our approach to conservation and environmental protection. By understanding the delicate balance of prehistoric ecosystems, we gain valuable insights into the importance of preserving biodiversity in our own time.

The final chapter on dinosaurs may have been written millions of years ago, but their story continues to inspire scientific curiosity, spark our imaginations, and serve as a powerful reminder of the remarkable history of life on Earth.

Bibliography and Further Reading

Bakker, Robert T. (1986). *The Dinosaur Heresies: New Theories on the Origins and Evolution of Dinosaurs.* Penguin Random House.

Benton, Michael J. (2004). *Vertebrate Paleontology (Third Edition):* Blackwell Publishing.

Brusatte, Steve (2012). *Dinosaur Paleobiology*: Johns Hopkins University Press.

Holtz Jr., Thomas R. (2000). *Dinosaurs: A Concise Natural History.* University of California Press.

Paul, Gregory S. (2002). *Dinosaurs of the Air: The Evolution and Extinction of Pterosaurs.* Yale University Press.

Sampson, Stephen D., & Witmer, Lawrence M. (2007). *The Dinosaur Way: An Exhibition of Skeletal mounts, Life-Size Replicas, Dioramas, and Interactive Exhibits.* Indiana University Press.

Journal of Paleontology (https://www.jstor.org/journal/jpaleontology)

PLOS One (https://plos.org/)

Proceedings of the Royal Society B: Biological Sciences (https://royalsocietypublishing.org/journal/rspb)

Further Reading:

Smithsonian National Museum of Natural History: Dinosaur Hall
https://naturalhistory.si.edu/exhibits

American Museum of Natural History: Dinosaurs
https://www.amnh.org/dinosaurs/the-dinosaurs-on-display

Natural History Museum, London: Dinosaurs
https://www.nhm.ac.uk/discover/dinosaurs.html

Planet Dinosaur (2011) - BBC Studios

Prehistoric Planet (2023) - Apple TV+

Walking with Dinosaurs (1999) - BBC Studios

www.ingramcontent.com/pod-product-compliance
Lightning Source LLC
Chambersburg PA
CBHW070246230526
45470CB00002B/499